Multivariate Analysis

Beyond the Surface

Preface

In the age of big data, we are constantly inundated with an overwhelming amount of information that can be challenging to interpret and make sense of. Multivariate analysis offers a powerful tool to help us understand complex data structures and identify patterns and relationships that might otherwise go unnoticed.

This book, Multivariate Analysis Beyond the Surface, is a comprehensive guide to the world of multivariate analysis. It takes the reader on a journey from the basic concepts and techniques to the more advanced and sophisticated methods used in real-world applications.

 One of the strengths of this book is its emphasis on going beyond the surface of the data to uncover the hidden relationships and patterns that lie beneath. It offers a range of advanced techniques such as factor analysis, cluster analysis, discriminant analysis, and principal component analysis that enable the reader to explore the data in greater depth and gain deeper insights.

Whether you are a student, a researcher, or a practitioner, this book will prove to be an invaluable resource. It is a must-read for anyone who wants to master the art of multivariate analysis and unlock the full potential of their data. I highly recommend this book to anyone who is interested in exploring the fascinating world of multivariate analysis beyond the surface.

Table of Contents

Chapter 1: Introduction to multivariate analysis

Multivariate analysis refers to a collection of statistical techniques used to analyze and model data with multiple variables. These techniques are used to gain insight into the relationships between variables and to make predictions about the values of these variables. Multivariate analysis is widely used in many fields, including psychology, economics, biology, and engineering.

Definition of multivariate analysis

Multivariate analysis involves analyzing data with more than one variable. These variables can be continuous or categorical. In statistics, variables are characteristics or properties that we measure or observe. For example, in a study about the height of individuals, height would be a variable.

Continuous variables are variables that can take any numerical value within a certain range. For example, height is a continuous variable because people can be any height between 0 and infinity (although realistically, there is a range of typical heights for humans). Other examples of continuous variables include weight, age, and temperature.

Categorical variables, on the other hand, are variables that take on a limited number of values that represent different categories or groups. For example, gender is a categorical variable because people can only be male or female (although there are many other gender identities that can be included as well). Other examples of categorical variables include race, education level, and favorite color.

So when we say that multivariate analysis can involve both continuous and categorical variables, it means that we can analyze relationships and patterns among variables that have numerical values that can be any number within a range (continuous), as well as variables that take on a limited number of values that represent different categories or groups (categorical).

The analysis can be used to explore the relationships between variables, identify patterns in the data, and make predictions about future observations.

Multivariate analysis techniques can be used to answer questions such as:

- Are there patterns in the data that can be used to group individuals or observations?
- Which variables are most closely related?
- Can we predict the value of one variable based on the values of others?
- How do different variables interact with each other?

1.1 Brief overview of the history and development of multivariate analysis

The development of multivariate analysis dates back to the early 20th century when the first techniques for analyzing data with multiple variables were introduced. One of the earliest and most influential techniques was principal component analysis, developed in the early

1900s by British mathematician Karl Pearson. This technique was used to reduce the dimensionality of data by identifying key underlying factors.

Over the next few decades, other techniques were developed to analyze multivariate data, including factor analysis, discriminant analysis, and cluster analysis. These techniques were used in various fields, including psychology, economics, and biology. With the advent of computers in the 1960s and 1970s, multivariate analysis became much more widely used, as it was now possible to analyze large data sets and perform complex calculations with ease.

Today, multivariate analysis continues to evolve and improve, with new techniques constantly being developed and refined. These techniques are used to analyze data in a wide range of fields, from finance and marketing to health care and environmental science.

1.2 Importance of multivariate analysis in modern data analysis

Multivariate analysis is an essential tool for modern data analysis as it allows us to better understand complex data sets. With the increasing availability of data from a wide variety of sources, including social media, sensor networks and electronic health records, multivariate analysis is becoming more important than ever.

Multivariate analysis is particularly important for:

- Identify patterns and relationships in large data sets
- Modeling complex systems with multiple interacting variables
- Make predictions based on multiple variables
- Developing new statistical models and techniques for analyzing data

As the volume and complexity of data continues to grow, the need for advanced multivariate analysis techniques will only increase. Whether you're a researcher, business analyst, or data scientist, an understanding of multivariate analysis is essential for success in today's data-driven world.

Chapter 2: Multivariate Distributions and Descriptive Statistics

Multivariate distributions and descriptive statistics are fundamental concepts in multivariate analysis. They provide a way to summarize and describe the relationships between multiple variables in a dataset. In this chapter, we will discuss the different types of multivariate distributions, as well as the various descriptive statistics and measures of association that are commonly used in multivariate analysis.

2.1 Overview of Multivariate Distributions

In multivariate analysis, a distribution refers to the set of possible values for a collection of variables. Multivariate distributions can be divided into two broad categories: continuous and discrete. Continuous multivariate distributions are characterized by a probability density function, while discrete multivariate distributions are characterized by a probability mass function.

One of the most commonly used continuous multivariate distributions is the multivariate normal distribution, which is characterized by its mean vector and covariance matrix. Imagine you have a dataset that includes the heights and weights of a group of people. You can represent this data as a multivariate normal distribution, where the mean vector represents the average height and weight of the group, and the covariance matrix represents the relationship between the two variables. So, if the covariance between height and weight is positive, it means that taller people tend to weigh more, and vice versa. This information can be used to make predictions about the height and weight of new people based on the patterns observed in the dataset.

The multivariate normal distribution is a useful distribution because it can be used to model many types of data that occur in practice, such as measurements of physical quantities, financial data, and biological data.

Examples:

- Physical measurements: Suppose we are measuring the height, weight, and age of a group of people. We can model these measurements as a multivariate normal distribution, where the mean vector represents the average height, weight, and age of the population, and the covariance matrix represents the variability and correlations between these variables.

- Financial data: Suppose we are analyzing the returns of a portfolio of stocks over a period of time. We can model these returns as a multivariate normal distribution, where the mean vector represents the average return of each stock, and the covariance matrix represents the variability and correlations between these returns.

- Biological data: Suppose we are analyzing gene expression data for a group of cells. We can model the expression levels of multiple genes as a multivariate normal distribution, where the mean vector represents the average expression level of each gene, and the covariance matrix represents the variability and correlations between these expression levels.

In addition to the multivariate normal distribution, there are many other types of continuous multivariate distributions, including the multivariate t-distribution, the multivariate gamma distribution, and the multivariate logistic distribution. Each of these distributions has its own set of properties and is suited to modeling different types of data.

The multivariate t-distribution is a continuous multivariate distribution that is similar to the normal distribution, but is more flexible in its ability to model data with heavier tails (i.e., more extreme values). It is often used in finance to model stock returns, which are known to have fat tails and skewness.

The multivariate gamma distribution is a continuous multivariate distribution that is used to model data that are non-negative and skewed. This distribution is commonly used in insurance to model the frequency and severity of claims, which are often non-negative and exhibit skewness.

The multivariate logistic distribution is a continuous multivariate distribution that is used to model data that are bounded between 0 and 1, such as proportions and probabilities. It is commonly used in epidemiology to model the prevalence and incidence of diseases, which are often expressed as proportions or probabilities.

Discrete multivariate distributions are often used to model data that take on only a finite number of values. This means that the data can be divided into categories or groups, and each observation falls into one of these categories.

For example, in genetics, researchers may be interested in studying the frequencies of different alleles (versions of a gene) in a population. The data collected would be discrete, as there are only a finite number of possible alleles. The multinomial distribution is commonly used to model this type of data, as it describes the probabilities of observing different combinations of categories.

Another example is in economics, where researchers may be interested in modeling the number of people in different income brackets. The data collected would also be discrete, as there are only a finite number of possible income brackets. The Poisson distribution can be used to model this type of data, as it describes the probability of observing a certain number of events (in this case, people in a specific income bracket) over a fixed interval of time or space.

Finally, in social sciences, researchers may be interested in modeling the number of successes in a fixed number of trials, such as the number of correct answers on a test. This type of data can also be modeled using a discrete distribution, such as the binomial distribution. The binomial distribution describes the probabilities of observing a certain number of successes in a fixed number of trials, given a specific probability of success for each trial.

2.2 Multivariate Descriptive Statistics

In addition to understanding the underlying distribution of a dataset, it is important to summarize and describe the data using descriptive statistics. Multivariate descriptive statistics are used to describe the relationships between multiple variables in a dataset.

2.2.1 Mean vector

One commonly used descriptive statistic is the mean vector, which is the vector of means for each variable in the dataset. The mean vector provides information about the central tendency of the data and is often used as a baseline for comparison with other statistics.

Suppose we have a dataset that includes the heights and weights of a sample of 100 people. We want to use multivariate analysis to explore the relationship between height and weight.

To begin our analysis, we can calculate the mean vector for the dataset. The mean vector will give us the average height and weight for our sample. Let's say that our mean vector is (68 inches, 150 pounds). This means that the average height of our sample is 68 inches and the average weight is 150 pounds.

We can use this mean vector as a baseline for comparison with other statistics. For example, if we calculate the mean vector for a different sample of 100 people and find that the mean height is 70 inches and the mean weight is 155 pounds, we can conclude that this sample is taller and heavier than our original sample.

Similarly, we can use the mean vector to identify any outliers in our dataset. If we find that one person in our sample has a height of 80 inches and a weight of 250 pounds, we can compare these values to the mean vector to see that they are much higher than the average for our sample. This could indicate that this person is an outlier and should be further investigated.

2.2.2 Covariance matrix

Another useful descriptive statistic is the covariance matrix, which provides information about the variability and the relationship between the variables in the dataset. The covariance matrix can be used to identify which variables are most strongly correlated with each other, and to detect patterns and trends in the data.

Let's say we have a dataset containing the scores of 50 students on three different exams: math, science, and history. The dataset can be represented as a matrix, where each row represents a student and each column represents a different exam. Here is an example of what the dataset might look like:

	Math	Science	History
1	75	80	90
2	80	85	95
3	70	75	80
...	...	...	...
50	85	90	95

To calculate the covariance matrix, we need to first calculate the mean of each variable. We can do this by adding up all the scores for each exam and dividing by the total number of students. Here are the mean scores for each exam:

- Mean(Math) = (75+80+70+...+85)/50 = 80

- Mean(Science) = (80+85+75+...+90)/50 = 85

- Mean(History) = (90+95+80+...+95)/50 = 90

Next, we need to calculate the variance and covariance between each pair of variables. The variance of a variable measures how much the scores of that variable vary around its mean. The covariance between two variables measures how much they vary together. Here are the formulas for the variance and covariance:

- $Var(X) = (1/n) * \Sigma(x\_i - \mu\_X)^2$

- $Cov(X,Y) = (1/n) * \Sigma(x\_i - \mu\_X)(y\_i - \mu\_Y)$

where X and Y are two different variables, n is the number of observations, x_i and y_i are the individual scores, and μ_X and μ_Y are the means of X and Y, respectively.

Using these formulas, we can calculate the variance and covariance for our example dataset:

- $Var(Math) = (1/50) * \Sigma(x\_i - 80)^2 = 50$

- $Var(Science) = (1/50) * \Sigma(y\_i - 85)^2 = 62.5$

- $Var(History) = (1/50) * \Sigma(z\_i - 90)^2 = 62.5$

- Cov(Math, Science) = (1/50) * Σ(x_i - 80)(y_i - 85) = 56.25

- Cov(Math, History) = (1/50) * Σ(x_i - 80)(z_i - 90) = 62.5

- Cov(Science, History) = (1/50) * Σ(y_i - 85)(z_i - 90) = 68.75

Using these values, we can construct the covariance matrix:

	Math	Science	History
Math	50	56.25	62.5
Science	56.25	62.5	68.75
History	62.5	68.75	62.5

The diagonal elements of the matrix represent the variances of each variable. For example, the variance of Math is 25.00, the variance of Science is 36.00, and the variance of English is 36.00.

The off-diagonal elements represent the covariances between pairs of variables. For example, the covariance between Math and Science is 56.25, which indicates that there is a positive correlation between the scores on these two exams. Similarly, the covariance between Math and English is 45.00, while the covariance between Science and English is 27.00.

By analyzing the covariance matrix, we can understand the relationships between the variables in our dataset. In this example, we see that there is a strong positive correlation between Math and Science scores, while the correlation between Science and English is weaker. The covariance matrix can also be used to detect patterns and trends in the data, which can help us make informed decisions and predictions.

2.3 Measures of Association and Dependence

In addition to descriptive statistics, multivariate analysis often involves measures of association and dependence, which are used to quantify the relationship between two or more variables in a dataset. There are some commonly used measures of association and dependence.

2.3.1 Correlation coefficient

The correlation coefficient is a statistical measure used to determine the strength and direction of the linear relationship between two variables. It is commonly denoted by the symbol "r," and its value ranges from -1 to +1.

If the correlation coefficient is close to +1, this indicates a strong positive linear relationship between the two variables, meaning that when one variable increases, the other variable tends to increase as well. On the other hand, if the correlation coefficient is close to -1, this indicates a strong negative linear relationship, meaning that when one variable increases, the other variable tends to decrease. If the correlation coefficient is close to 0, this indicates a weak or no linear relationship between the two variables.

For example, let's say we want to investigate the relationship between the number of hours studied and the score received on an exam. We collect data from 50 students, recording the number of hours they studied and the score they received on the exam. We can then use the correlation coefficient to determine the strength and direction of the relationship between these two variables.

After analyzing the data, we find that the correlation coefficient between the number of hours studied and the exam score is 0.75. This indicates a strong positive linear relationship between the two variables. In other words, students who study more tend to receive higher scores on the exam.

However, it's important to note that correlation does not necessarily imply causation. Just because there is a strong positive correlation between the number of hours studied and the exam score does not mean that studying more directly causes a higher exam score. There could be other factors at play, such as natural ability or motivation.

2.3.2 Covariance

Covariance is a statistical measure that indicates the degree to which two variables vary together. It is calculated by multiplying the deviation of each data point from the mean of their respective variable and then taking the average of those products. A positive covariance value indicates that the two variables tend to increase or decrease together, while a negative covariance value indicates that they tend to vary in opposite directions.

To illustrate the concept of covariance, let's consider an example.

Suppose we have data on the number of hours worked and the salary earned by five employees in a company, as shown in the table below:

Employee	Hours Worked (x)	Salary Earned (y)
1	20	35000
2	15	27000
3	25	42000
4	18	32000
5	22	38000

We can calculate the covariance between these two variables using the following formula:

$$\text{Covariance} = \Sigma(x - \bar{x})(y - \bar{y}) / n\text{-}1$$

Where x and y are the respective variables (hours worked and salary earned), $\bar{x}$ and $\bar{y}$ are their respective means, and n is the number of observations.

Using the formula, we can compute the covariance between hours worked and salary earned as follows:

Covariance = [(20-20.0)*(35000-34800) + (15-20.0)*(27000-34800) + (25-20.0)*(42000-34800) + (18-20.0)*(32000-34800) + (22-20.0)*(38000-34800)] / (5-1)

Covariance = -1600

The negative value of covariance (-1600) indicates that there is a negative relationship between the two variables. In other words, employees who work more hours tend to earn lower salaries, and those who work fewer hours tend to earn higher salaries.

It's important to note that covariance alone doesn't give us a complete picture of the relationship between two variables. For example, in our example, we can see that there is a negative covariance between hours worked and salary earned. However, we don't know if the relationship is strong or weak, and we don't know the exact direction of the relationship (e.g., is it linear, quadratic, etc.). To gain a more complete understanding of the relationship between the two variables, we would need to use other statistical measures, such as correlation coefficient or regression analysis.

2.3.3 Regression analysis

Regression analysis is a statistical technique used to analyze the relationship between a dependent variable and one or more independent variables. The dependent variable is the variable that we want to predict or explain, while the independent variables are the variables that we believe have an influence on the dependent variable. Regression analysis can help us understand how changes in the independent variables affect the dependent variable.

To illustrate the concept of regression analysis, let's consider an example. Suppose we want to analyze the relationship between the price of a house (dependent variable) and its size (independent variable). We collect data from 20 houses in a neighborhood, recording the size of each house (in square feet) and its price (in thousands of dollars). The data is shown in the table below:

House	Size (sq. ft.)	Price (in thousands)
1	1600	280
2	1260	200
3	1780	310
4	1300	215
5	1560	285
6	1530	290
7	1450	255
8	1800	315
9	1100	200
10	1650	300
11	1290	240
12	1850	325
13	1220	215
14	2000	400
15	1180	210
16	1420	280
17	1580	310
18	1150	200
19	1650	295
20	1400	260

We can use regression analysis to find the equation of a line that best fits the data and predicts the price of a house based on its size. This line is called the regression line or the line of best fit.

There are different types of regression analysis, but the most commonly used method is linear regression. In linear regression, we assume that there is a linear relationship between the dependent variable and the independent variable(s). The equation of the regression line can be written as:

$$y = b0 + b1*x$$

Where y is the dependent variable (price of the house), x is the independent variable (size of the house), b0 is the intercept, and b1 is the slope of the line.

We can use statistical software or spreadsheet programs to calculate the regression line. For our example, the regression line is:

$$Price = -33.9 + 0.164 * Size$$

The intercept (b0) is -33.9, which means that a house with a size of 0 square feet has an estimated price of -33.9 thousands of dollars. Of course, this doesn't make sense in the real world, so we only use the line to predict prices for houses with sizes within the range of the data we collected.

The slope (b1) is a key parameter in linear regression analysis, as it tells us how much the dependent variable changes for every unit increase in the independent variable. In the example of the relationship between the price of a house and its size, the slope of the regression line is 0.164.

This means that for every 1 square foot increase in the size of a house, the estimated price increases by 0.164 thousand dollars (or $164). This is a positive relationship, which means that as the size of the house increases, the price of the house also tends to increase. The slope value of 0.164 indicates that the relationship is not particularly strong, as a change of 1 square foot only leads to a change of $164 in price.

It's important to note that the slope only represents the relationship between the two variables within the range of data that was collected. It's possible that the relationship between the size of a house and its price could be different outside of this range, or that there are other variables that also affect the price of a house.

The slope can also be used to make predictions about the dependent variable based on the value of the independent variable. For example, if we want to predict the price of a house with a size of 1500 square feet, we can use the equation of the regression line:

Price = -33.9 + 0.164 * Size

Plugging in the value of 1500 for Size, we get:

Price = -33.9 + 0.164 * 1500

Price = $230.1 thousand

This means that we would predict the price of a house with a size of 1500 square feet to be $230.1 thousand based on the relationship between size and price observed in the data we collected. However, it's important to keep in mind that there may be other factors that affect the price of a house, and that this prediction is only an estimate based on the available data.

2.3.4 Canonical correlation analysis

Canonical correlation analysis is a statistical technique used to explore the relationship between two sets of variables. Specifically, it measures the correlation between linear combinations of the two sets of variables, called canonical variates. The goal of canonical correlation analysis is to identify the strongest linear relationship between the two sets of variables and to provide insight into how they are related.

Let's consider an example to better understand canonical correlation analysis. Suppose we are interested in understanding the relationship between a person's level of physical activity and their overall health. We have collected data on two sets of variables: the first set includes measures of physical activity, such as the number of steps taken per day, the amount of time spent exercising, and the number of days per week that the person engages in physical activity. The second set includes measures of overall health, such as blood pressure, cholesterol levels, body mass index (BMI), and self-reported health status.

To perform canonical correlation analysis, we first need to standardize the two sets of variables so that they have a mean of zero and a standard deviation of one. This is necessary because the technique is sensitive to differences in the scale of the variables. We then calculate the correlation between all possible linear combinations of the two sets of variables. The resulting correlations are called canonical correlations, and they range in value from zero to one.

The output of canonical correlation analysis includes the canonical correlations themselves, as well as the canonical weights, which are used to calculate the canonical variates. The canonical weights are similar to regression coefficients and represent the contribution of each variable to the canonical variate. The canonical variates are the linear combinations of the original variables that have the highest correlation with each other.

In our example, let's say that the first two canonical correlations are both statistically significant. The first canonical correlation is 0.8, indicating a strong relationship between the two sets of variables. The second canonical correlation is 0.6, indicating a weaker relationship. We can also look at the canonical weights to understand which variables are contributing the most to each canonical variate.

Suppose that the first canonical variate is most strongly related to the number of steps taken per day, the amount of time spent exercising, and self-reported health status. This suggests that people who take more steps per day, spend more time exercising, and report better health are likely to have higher scores on the first canonical variate. The second canonical variate might be most strongly related to BMI and cholesterol levels, indicating that people with higher BMIs and cholesterol levels may have lower scores on this variate.

These measures of association and dependence are used to identify relationships between variables and to make predictions about the values of one variable based on the values of others.

2.4 Conclusion chapter 2

Multivariate distributions and descriptive statistics provide a foundation for multivariate analysis. They allow us to summarize and describe the relationships between multiple variables in a dataset, and to make predictions based on those relationships. Measures of association and dependence allow us to quantify those relationships and to identify patterns and trends in the data. Understanding these concepts is essential for any multivariate analysis, and provides a solid foundation for more advanced techniques.

Chapter 3: Multivariate Inference

Multivariate inference is a critical component of multivariate analysis. It involves making statistical inferences about the population parameters based on a sample of data. In this chapter, we will discuss the different types of hypothesis testing, confidence intervals, and nonparametric tests used in multivariate analysis.

3.1 Hypothesis Testing in Multivariate Analysis

Hypothesis testing is a common tool used in multivariate analysis to test whether a sample of data is consistent with a given hypothesis about the population. In multivariate hypothesis testing, the null hypothesis often involves a statement about the population mean vector or covariance matrix.

The most commonly used hypothesis tests in multivariate analysis are the Hotelling's T2 test and the MANOVA (multivariate analysis of variance) test. Hotelling's T2 test is used to test the hypothesis that the mean vector of a population is equal to a given value. MANOVA, on the other hand, is used to test the hypothesis that the mean vectors of two or more populations are equal.

3.1.1 Hotelling's T2 test

Hotelling's T2 test is a statistical technique used to test the hypothesis that the mean vector of a population is equal to a given value. The mean vector is a vector that contains the means of multiple variables for a given population. The test is commonly used in multivariate analysis, where multiple variables are analyzed simultaneously.

To understand Hotelling's T2 test, let's consider an example. Suppose we are interested in comparing the performance of two different companies on multiple financial measures, such as revenue, profit margin, and return on investment. We collect data on these variables for each company and want to test whether there is a significant difference in the mean vector of the two populations.

The first step is to calculate the sample mean vector for each company. This involves taking the mean of each variable across all observations for that company. We can then calculate the pooled sample mean vector, which is the average of the two sample mean vectors.

Next, we calculate the sample covariance matrix for each company. The sample covariance matrix is a matrix that contains the covariance between all pairs of variables. We can then calculate the pooled sample covariance matrix, which is a weighted average of the two sample covariance matrices.

Using the sample mean vector and covariance matrix, we can calculate Hotelling's T2 statistic. This is a measure of the distance between the two mean vectors, standardized by the covariance matrix. It is calculated as:

$$T2 = n_1 n_2 / (n_1 + n_2) * (x_1 - x_2)' S^{-1} (x_1 - x_2)$$

where n1 and n2 are the sample sizes for the two companies, $\bar{x}1$ and $\bar{x}2$ are the sample mean vectors for each company, and S^{-1} is the inverse of the pooled sample covariance matrix.

The T2 statistic follows a chi-square distribution with p degrees of freedom, where p is the number of variables being analyzed. We can calculate the critical value of T2 at a given significance level to determine whether the observed T2 statistic is statistically significant. If the observed T2 statistic is greater than the critical value, we reject the null hypothesis and conclude that there is a significant difference in the mean vector of the two populations.

In our example, if the T2 statistic is found to be statistically significant, we can conclude that there is a significant difference in the financial performance of the two companies. We can then investigate which variables are driving the difference and develop strategies to improve performance in those areas.

3.1.2 MANOVA

Multivariate Analysis of Variance (MANOVA) is a statistical technique used to test the hypothesis that the mean vectors of two or more populations are equal. MANOVA is similar to Analysis of Variance (ANOVA), but instead of analyzing a single dependent variable, it considers multiple dependent variables simultaneously.

To understand MANOVA, let's consider an example. Suppose we are interested in comparing the academic performance of three different schools. We collect data on three different subjects - Mathematics, Science, and English - for each school. Our hypothesis is that the mean academic performance of the three schools is equal.

The first step in MANOVA is to calculate the mean vector and covariance matrix for each school for the three subjects. This involves taking the mean of each subject across all observations for that school and calculating the covariance between all pairs of subjects for that school. We can then calculate the pooled sample mean vector and covariance matrix for the three schools.

Using the sample mean vector and covariance matrix, we can calculate the MANOVA F-statistic. This is a measure of the distance between the mean vectors of the three schools, standardized by the covariance matrix. It is calculated as:

$$F = ((n - g)p / (g - 1)(n - p - 1)) * |W|^{-1} * |B|$$

where n is the total sample size, g is the number of groups (in this case, three schools), p is the number of dependent variables (in this case, three subjects), W is the pooled sample covariance matrix, and B is the between-groups covariance matrix.

The F-statistic follows an F-distribution with (g - 1) and (n - p - 1) degrees of freedom. We can calculate the critical value of F at a given significance level to determine whether the observed F-statistic is statistically significant. If the observed F-statistic is greater than the critical value, we reject the null hypothesis and conclude that there is a significant difference in the mean vectors of the three schools.

In our example, if the MANOVA F-statistic is found to be statistically significant, we can conclude that there is a significant difference in the academic performance of the three schools. We can then investigate which subjects are driving the difference and develop strategies to improve performance in those areas.

3.2 Confidence Intervals for Multivariate Means and Covariance Matrices

Confidence intervals provide a way to estimate the range of plausible values for a population parameter based on a sample of data. In multivariate analysis, confidence intervals are often used to estimate the mean vector and covariance matrix of a population.

The most commonly used confidence interval for the mean vector is the Hotelling's T2 confidence interval (see previous chapter). This interval provides an estimate of the range of plausible values for the population mean vector based on the sample mean vector and covariance matrix.

3.2.1 Wishart Distribution

For the covariance matrix, a confidence interval can be constructed using the Wishart distribution. The Wishart distribution is a probability distribution that describes the distribution of the sample covariance matrix for a multivariate normal distribution. It is characterized by two parameters: the degrees of freedom and the scale matrix.

To construct a confidence interval for the covariance matrix, we start by estimating the sample covariance matrix from our data. We then calculate the Wishart distribution using the estimated covariance matrix and the degrees of freedom parameter. The degrees of freedom parameter is typically chosen based on the size of the sample and the number of variables. As the sample size and number of variables increase, the degrees of freedom parameter should also increase.

Once we have calculated the Wishart distribution, we can use it to construct a confidence interval for the true population covariance matrix. The confidence interval is typically constructed using the percentiles of the Wishart distribution. For example, a 95% confidence interval for the covariance matrix would include the 2.5th and 97.5th percentiles of the Wishart distribution.

To illustrate this process, let's consider an example. Suppose we have data on the heights and weights of a sample of individuals. We want to estimate the covariance matrix for these two variables and construct a 95% confidence interval for the true population covariance matrix.

We start by estimating the sample covariance matrix from our data. Let X be a matrix where the rows represent individuals and the columns represent variables (height and weight). We calculate the sample covariance matrix as:

$$S = (1 / (n - 1)) * (X - mu)' * (X - mu)$$

where mu is the mean vector of X and n is the sample size.

Next, we calculate the Wishart distribution using the estimated covariance matrix and the degrees of freedom parameter. Let's say we choose the degrees of freedom parameter to be 10. We calculate the Wishart distribution as:

$$W \sim \text{Wishart}(S, 10)$$

We can then use the percentiles of the Wishart distribution to construct a 95% confidence interval for the true population covariance matrix. Let's say the 2.5th and 97.5th percentiles of the Wishart distribution are 0.20 and 0.50, respectively. This means that we can be 95% confident that the true population covariance matrix falls within the interval [0.20, 0.50].

3.3 Nonparametric Tests for Multivariate Data

Nonparametric tests are statistical tests that do not assume a specific distribution for the data. In multivariate analysis, nonparametric tests are often used when the assumptions of parametric tests are not met.

3.3.1 Wilks' Lambda Test

One commonly used nonparametric test in multivariate analysis is the Wilks' lambda test. Wilks' lambda test is a nonparametric test that uses the Wilks' lambda statistic to test the hypothesis of equality of mean vectors. The test is based on the assumption that the populations are multivariate normal and have possibly different covariance matrices. The null hypothesis is that the mean vectors of all populations are equal.

The Wilks' lambda statistic is defined as the ratio of the determinant of the pooled within-group covariance matrix to the determinant of the pooled total covariance matrix. The test involves comparing the value of Wilks' lambda statistic to a critical value obtained from the F distribution. The degrees of freedom for the F distribution are equal to the number of populations minus one and the number of variables.

To illustrate this test, let's consider an example. Suppose we want to compare the means of three groups with respect to three variables: X, Y, and Z. We assume that the populations are multivariate normal, but we do not assume that the covariance matrices are equal. We collect data from each of the three groups and calculate the sample mean vectors and sample covariance matrices.

We can then calculate the Wilks' lambda statistic as follows:

- Calculate the within-group covariance matrices for each group.
- Calculate the pooled within-group covariance matrix by averaging the within-group covariance matrices.
- Calculate the total pooled covariance matrix by averaging the covariance matrices of all groups.
- Calculate the Wilks' lambda statistic as the ratio of the determinant of the pooled within-group covariance matrix to the determinant of the pooled total covariance matrix.

We can then use the calculated Wilks' lambda statistic to test the null hypothesis that the mean vectors of all populations are equal. The test involves comparing the Wilks' lambda statistic to a critical value obtained from the F distribution with degrees of freedom equal to (number of populations - 1) and (number of variables). If the calculated Wilks' lambda statistic is less than the critical value, we reject the null hypothesis and conclude that at least one population mean vector is different from the others.

3.3.2 The Permutation Test

Another commonly used nonparametric test is the permutation test. The permutation test is a nonparametric statistical test that can be used to test a null hypothesis when the assumptions of traditional parametric tests are not met. This test involves randomly permuting the labels of the observations and calculating the test statistic for each permutation. The distribution of the test statistics under the null hypothesis is then approximated using these permutations, and a p-value is calculated based on how extreme the observed test statistic is compared to this distribution.

To illustrate the permutation test, let's consider an example. Suppose we have a dataset consisting of two groups: a control group and a treatment group. The outcome variable of interest is a continuous variable, such as blood pressure. We want to test the hypothesis that the mean blood pressure in the treatment group is different from that in the control group.

A traditional t-test can be used to test this hypothesis, but it assumes that the data are normally distributed and that the variances of the two groups are equal. However, if these assumptions are not met, the t-test may not be appropriate. In this case, the permutation test can be used instead.

To perform the permutation test, we first calculate the observed test statistic. In this case, we can use the difference in means between the treatment group and the control group as our test statistic. We then randomly permute the labels of the observations, re-calculate the test statistic for each permutation, and store the value.

We repeat this process many times, generating a distribution of the test statistics under the null hypothesis. We can then calculate a p-value based on how extreme the observed test statistic is compared to this null distribution. If the p-value is less than our chosen significance level, we reject the null hypothesis and conclude that there is evidence of a difference in mean blood pressure between the treatment group and the control group.

The permutation test is useful because it makes few assumptions about the data and can be used with any type of outcome variable, including categorical variables. However, it can be computationally intensive and may not be appropriate for very large datasets.

3.4 Conclusion chapter 3

Multivariate inference is a critical component of multivariate analysis. Hypothesis testing, confidence intervals, and nonparametric tests provide a way to make statistical inferences about the population parameters based on a sample of data. Understanding these techniques is essential for drawing meaningful conclusions from multivariate data, and for making informed decisions based on the analysis.

Chapter 4: Multivariate Regression

Multivariate regression is a powerful technique in multivariate analysis that allows us to model the relationships between multiple independent variables and a dependent variable. In this chapter, we will discuss the different types of multivariate regression, including multiple regression, logistic regression, canonical correlation analysis, and discriminant analysis.

4.1 Multiple Regression

Multiple regression is a widely used statistical technique that examines the relationship between a dependent variable and two or more independent variables. It is commonly used in social and behavioral sciences, finance, and other fields where researchers want to predict the value of a dependent variable based on multiple predictors.

To illustrate multiple regression, let's consider an example where we want to predict a person's salary based on their level of education, years of work experience, and gender. In this example, salary is the dependent variable, and education, work experience, and gender are the independent variables.

To perform multiple regression, we start by collecting data on these variables for a sample of individuals. We then use a statistical software package to fit a regression model to the data. The model takes the form of:

$$Salary = b0 + b1Education + b2Experience + b3*Gender$$

In this equation, b0 is the intercept, or the value of the dependent variable when all the independent variables are equal to zero. The b1, b2, and b3 coefficients represent the estimated effect of each independent variable on the dependent variable, holding all other independent variables constant.

We can use the coefficients from the regression model to make predictions about a person's salary based on their level of education, years of work experience, and gender. For example, we might use the model to predict the salary of a person with a Bachelor's degree, 5 years of work experience, and who identifies as female.

In addition to estimating the coefficients, multiple regression also provides information about the overall fit of the model to the data. One commonly used metric is R-squared, which measures the proportion of variance in the dependent variable that can be explained by the independent variables. A high R-squared value indicates a good fit between the model and the data, while a low value indicates that the model is not a good fit.

Multiple regression can be used to test hypotheses about the relationship between the independent variables and the dependent variable. For example, we might want to test the hypothesis that there is no relationship between gender and salary, after controlling for education and work experience. This can be done using a hypothesis test on the b3 coefficient, which represents the effect of gender on salary.

4.2 Logistic Regression

Logistic regression is a statistical technique used to model the relationship between a binary dependent variable and one or more independent variables. The dependent variable in logistic regression is a categorical variable with two possible outcomes, typically represented as 0 or 1. The independent variables can be categorical or continuous.

To illustrate logistic regression, let's consider an example where we want to predict whether a customer will purchase a product or not, based on their age, income, and gender. In this example, the dependent variable is a binary variable that takes the value 1 if the customer purchased the product and 0 otherwise. The independent variables are age, income, and gender.

Logistic regression estimates the probability that the dependent variable will take on a particular value, given the values of the independent variables. The estimated probability is modeled as a function of the independent variables. In logistic regression, this function is typically modeled using the logistic or sigmoid function.

The logistic function takes the form:

$$P(y=1|x) = e^{\wedge}(b0 + b1x1 + b2x2 + \ldots + bnxn) / (1 + e^{\wedge}(b0 + b1x1 + b2x2 + \ldots + bnxn))$$

where $P(y=1|x)$ is the probability that the dependent variable takes on the value 1, given the values of the independent variables x1, x2, ..., xn. The b0, b1, b2, ..., bn coefficients represent the estimated effect of each independent variable on the log-odds of the dependent variable.

The log-odds of the dependent variable taking on the value 1, given the independent variables, can be expressed as:

$$\log(\text{odds}) = b0 + b1x1 + b2x2 + \ldots + bnxn$$

The odds of the dependent variable taking on the value 1 are equal to the probability of the dependent variable taking on the value 1, divided by the probability of the dependent variable taking on the value 0. The log-odds can take any value from negative infinity to positive infinity, and can be transformed into probabilities using the logistic function.

To estimate the coefficients in logistic regression, we use maximum likelihood estimation. This involves finding the values of the coefficients that maximize the likelihood of the observed data, given the model. The likelihood function is a function of the coefficients and represents the probability of observing the data given the model.

Once we have estimated the coefficients in the logistic regression model, we can use the model to make predictions about the probability of a customer purchasing the product, given their age, income, and gender. For example, we might use the model to predict the probability of a customer who is 30 years old, earns $50,000 per year, and identifies as female purchasing the product.

Logistic regression is widely used in fields such as epidemiology, biostatistics, and psychology. It is used to model relationships where the dependent variable takes on only two values, such as whether a patient will survive or die from a disease.

4.3 Canonical Correlation Analysis

Canonical correlation analysis (CCA) is a statistical technique used to explore the relationships between two sets of variables. It measures the linear association between two sets of variables and identifies the underlying patterns between them. It is a multivariate technique that can be used to analyze complex data and can be used in various fields such as biology, psychology, sociology, and economics.

For example, a researcher may be interested in studying the relationship between a set of physiological measures (such as heart rate, blood pressure, and cholesterol levels) and a set of psychological measures (such as anxiety levels, depression levels, and stress levels) in a group of patients. The researcher wants to investigate whether there is a relationship between the physiological measures and psychological measures, and if so, to what extent.

To perform CCA, the researcher will first collect data on both the physiological and psychological measures for the sample of patients. The data will be organized into two sets: the physiological measures set and the psychological measures set. Each set will contain multiple variables.

Next, the researcher will run the CCA to identify the relationship between the two sets of variables. The technique will identify the linear combinations of variables in each set that are most highly correlated with each other. These linear combinations are called canonical variables.

The researcher will then examine the correlations between the canonical variables. The correlation between the canonical variables indicates the strength of the relationship between the two sets of variables. If the correlation is high, it suggests that there is a strong relationship between the physiological and psychological measures. If the correlation is low, it suggests that there is little or no relationship between the two sets of variables.

The researcher can also perform hypothesis tests to determine whether the relationship between the two sets of variables is statistically significant. If the p-value is less than the significance level, the researcher can reject the null hypothesis and conclude that there is a significant relationship between the two sets of variables.

4.4 Discriminant Analysis

Discriminant analysis is a statistical technique that is commonly used to classify objects or individuals into groups based on their characteristics or attributes. It is a multivariate regression technique that can be used to model the relationships between a dependent variable and multiple independent variables. The goal of discriminant analysis is to identify the independent variables that best predict the value of the dependent variable.

For example, a researcher may be interested in studying the factors that predict whether a customer is likely to churn or not from a telecommunications company. The researcher may have collected data on various independent variables such as the customer's age, income, usage patterns, customer service experience, and so on. The dependent variable is whether the customer churned or not.

To perform discriminant analysis, the researcher will first collect data on the independent and dependent variables for the sample of customers. The data will be organized into two groups, one for customers who churned and the other for those who did not. Each group will contain multiple variables.

Next, the researcher will run the discriminant analysis to identify the independent variables that best predict the value of the dependent variable. The technique will identify the linear combinations of independent variables that are most effective in discriminating between the two groups.

The researcher will then use these linear combinations to create a discriminant function. The discriminant function can be used to predict whether a new customer is likely to churn or not based on their independent variable values. The researcher can also use the discriminant function to identify which independent variables have the most significant impact on the dependent variable.

Finally, the researcher will evaluate the accuracy of the discriminant function by using it to classify a new sample of customers as churners or non-churners. The accuracy of the classification can be measured using various metrics such as sensitivity, specificity, precision, and recall.

Discriminant analysis is widely used in fields such as psychology, sociology, and biology. It can be used to model relationships between multiple independent variables and a dependent variable, such as predicting whether a patient has a particular disease based on their symptoms.

4.5 Conclusion chapter 4

Multivariate regression is a powerful technique in multivariate analysis that allows us to model the relationships between multiple independent variables and a dependent variable. Multiple regression, logistic regression, canonical correlation analysis, and discriminant analysis are some of the commonly used techniques in multivariate regression. Understanding these techniques is essential for making informed decisions based on the analysis and drawing meaningful conclusions from multivariate data.

Chapter 5: Principal Component Analysis

Principal component analysis (PCA) is a widely used multivariate statistical technique that is used to reduce the complexity of multivariate data by identifying patterns and relationships among the variables. In this chapter, we will discuss the fundamentals of PCA, including the calculation of principal components, interpretation of principal components, and the applications of PCA.

5.1 Overview of Principal Component Analysis

Principal Component Analysis (PCA) is a statistical technique that reduces the complexity of high-dimensional datasets while retaining as much information as possible. PCA works by transforming the original dataset into a new coordinate system in which the new variables (i.e., principal components) are uncorrelated and ordered by their importance.

For example, suppose we have a dataset containing measurements of various physical characteristics of flowers, such as petal length, petal width, sepal length, and sepal width. These variables may be correlated, making it difficult to analyze the data. By applying PCA, we can identify the underlying patterns and relationships among these variables and reduce them to a smaller set of principal components.

PCA can be particularly useful in data visualization, where it can be challenging to visualize high-dimensional data. For instance, suppose we have a dataset of customer demographics, purchase histories, and product preferences for an e-commerce website. We can use PCA to reduce the number of variables and plot the data in two dimensions, making it easier to see clusters of customers with similar preferences and behaviors.

Moreover, PCA can be used to improve the performance of machine learning models by reducing the number of features that are fed into the model. For instance, suppose we have a dataset of images with high-resolution pixels. The high dimensionality of the dataset can make it difficult to train a machine learning model efficiently. We can use PCA to reduce the number of pixels by identifying the most important patterns in the images, resulting in a lower-dimensional dataset that is easier to process and less prone to overfitting.

5.2 Calculation of Principal Components

PCA involves a series of mathematical calculations to identify the principal components of a dataset. The first step is to calculate the covariance matrix of the original variables. The covariance matrix provides information on the degree of linear relationship between the variables.

Next, the eigenvalues and eigenvectors of the covariance matrix are calculated. The eigenvectors represent the directions of maximum variance in the dataset, while the eigenvalues represent the amount of variance explained by each eigenvector.

The first principal component is then identified as a linear combination of the original variables that explains the maximum amount of variance in the dataset. The weights of the original variables in the first principal component are determined by the eigenvector associated with the largest eigenvalue.

For example, suppose we have a dataset containing the heights and weights of individuals. The covariance matrix of the dataset would reveal that height and weight are positively correlated. The first principal component would be a linear combination of height and weight that explains the maximum amount of variation in the dataset.

The second principal component is then identified as a linear combination of the remaining variables that explains the maximum amount of variation in the dataset, subject to the constraint that it is uncorrelated with the first principal component. The weights of the original variables in the second principal component are determined by the eigenvector associated with the second largest eigenvalue.

For example, in the same height and weight dataset, the second principal component may be a linear combination of age and body fat percentage. This component would be uncorrelated with the first principal component and would explain the maximum amount of variation in the dataset that is not explained by the first principal component.

This process is repeated to identify additional principal components. The number of principal components retained depends on the amount of variation in the dataset that needs to be explained. Typically, principal components with eigenvalues greater than one are retained, as they explain more variance than a single original variable.

The eigenvalue and eigenvector of the correlation matrix of the original variables are used to calculate the principal components. The eigenvalues represent the amount of variance explained by each principal component, while the eigenvectors represent the weights of each variable in the principal component.

5.3 Interpretation of Principal Components

PCA provides a way to interpret the underlying patterns and relationships among the variables in a dataset. The principal components can be interpreted as new variables that are uncorrelated with each other and provide a simplified representation of the original variables.

The interpretation of principal components depends on the specific dataset being analyzed and the application in which it is being used. In some cases, the principal components may represent underlying factors or constructs that are not directly observable. These constructs may be related to psychological or social phenomena, such as intelligence or personality traits.

For example, suppose we have a dataset containing information on individuals' scores on a battery of cognitive tests. By applying PCA, we may identify a set of principal components that represent underlying factors, such as verbal ability, spatial reasoning, or working memory capacity. These factors may not be directly observable, but they can provide valuable insights into the underlying cognitive processes that are involved in performing well on the tests.

In other cases, the principal components may represent important patterns or relationships among the variables. For example, in financial data, PCA can be used to identify trends in stock prices or patterns in interest rates. By identifying these patterns, analysts can make informed decisions about investment strategies or risk management.

Similarly, in gene expression data, PCA can be used to identify patterns of gene expression that are related to disease states or other biological phenomena. These patterns can provide insights into the underlying mechanisms of disease and may lead to the development of new treatments or therapies.

It is important to note that the interpretation of principal components is not always straightforward and requires careful consideration of the specific context and the goals of the analysis. However, by identifying the most important patterns and relationships among the variables, PCA can provide valuable insights into complex datasets and help to simplify and interpret the data.

5.4 Applications of Principal Component Analysis

PCA is a versatile technique that has many applications in various fields. In finance, PCA is used to identify patterns in stock returns and to construct optimal portfolios. By identifying patterns in stock returns, investors can make more informed decisions about investment strategies and risk management. For example, PCA can be used to identify stocks that are likely to move together, allowing investors to construct portfolios that are diversified across different sectors or asset classes.

In biology, PCA is used to analyze gene expression data and to identify clusters of genes with similar expression patterns. This can provide insights into the underlying biological processes that are involved in disease states and may lead to the development of new treatments or therapies. For example, PCA can be used to identify patterns of gene expression that are associated with specific diseases, allowing researchers to develop targeted treatments that are tailored to the underlying biological mechanisms.

In psychology, PCA is used to identify underlying factors that contribute to personality traits and to study the relationships between different aspects of behavior. By identifying these underlying factors, researchers can better understand the mechanisms that contribute to personality and behavior, which may lead to the development of more effective interventions or treatments. For example, PCA can be used to identify the underlying factors that contribute to depression or anxiety, allowing researchers to develop more effective therapies that target these factors.

In addition to its analytical applications, PCA can also be used for data visualization. By projecting high-dimensional data onto a lower-dimensional space defined by the principal components, PCA can help to identify patterns and relationships in the data that are not apparent in the original high-dimensional space. For example, PCA can be used to visualize complex datasets such as brain imaging data or social networks, allowing researchers to identify patterns of activity or social interactions that are not visible in the original data.

5.5 Conclusion chapter 5

PCA is a powerful technique in multivariate analysis that allows us to identify underlying patterns and relationships among the variables in a dataset. By transforming a set of correlated variables into a set of uncorrelated principal components, PCA provides a simplified representation of complex datasets that can aid in interpretation and analysis. PCA has many applications in various fields and can be used to analyze and visualize high-dimensional data. Understanding the principles and applications of PCA is essential for making informed decisions based on multivariate data.

Chapter 6: Factor Analysis

Factor analysis is a multivariate statistical technique used to identify underlying latent variables, or factors, that explain the correlations among a set of observed variables. In this chapter, we will discuss the fundamentals of factor analysis, including the calculation of factors, interpretation of factors, and the applications of factor analysis.

6.1 Overview of Factor Analysis

Factor analysis is a statistical technique that is widely used in many fields, including psychology, sociology, and economics. The main goal of factor analysis is to identify the underlying constructs or factors that contribute to observed variables. These factors can represent a wide range of phenomena, including intelligence, personality traits, attitudes, or socio-economic status.

For example, in psychology, factor analysis is often used to study intelligence. Researchers may administer a battery of cognitive tests to participants and then use factor analysis to identify the underlying factors that contribute to the observed scores on the tests. These factors may represent different aspects of intelligence, such as verbal ability, spatial reasoning, or working memory capacity. By identifying these factors, researchers can gain insights into the underlying cognitive processes that are involved in performing well on the tests.

Similarly, in sociology, factor analysis is often used to study socio-economic status. Researchers may collect data on a range of variables, such as income, education, and occupation, and then use factor analysis to identify the underlying factors that contribute to these variables. These factors may represent different aspects of socio-economic status, such as wealth, education, or occupational prestige. By identifying these factors, researchers can gain insights into the social and economic factors that contribute to social inequality and stratification.

In economics, factor analysis is often used to study consumer behavior. Researchers may collect data on a range of variables, such as income, spending habits, and demographic characteristics, and then use factor analysis to identify the underlying factors that contribute to these variables. These factors may represent different aspects of consumer behavior, such as price sensitivity, brand loyalty, or demographic preferences. By identifying these factors, researchers can gain insights into the factors that influence consumer behavior and develop more effective marketing strategies.

Overall, factor analysis is a powerful technique that can help simplify complex datasets and identify the key factors that contribute to the observed patterns. By identifying these underlying factors, researchers can gain insights into the mechanisms that drive complex phenomena and develop more effective interventions or strategies.

6.2 Calculation of Factors

Factor analysis involves a series of mathematical calculations to identify the underlying factors that explain the correlations among the observed variables. The factors are calculated based on the variance-covariance matrix of the observed variables. The goal is to identify a set of factors that account for as much of the variance in the observed variables as possible.

The most common method for factor analysis is principal component analysis (PCA) (see previous chapter). The principal components represent the underlying factors that contribute to the observed variables. The factors are extracted based on the eigenvalues of the correlation matrix, which represent the amount of variance explained by each factor.

6.3 Interpretation of Factors

The interpretation of the factors in factor analysis is highly dependent on the specific dataset being analyzed and the application. The factors can be interpreted as underlying constructs that contribute to the observed variables. For example, in a study of personality traits, the factors might represent different dimensions of personality, such as extraversion, agreeableness, or neuroticism. In a study of socio-economic status, the factors might represent different aspects of socio-economic status, such as education, occupation, or income.

However, the interpretation of the factors is not always straightforward. Factors can be complex and difficult to interpret, and the relationships between the factors and the observed variables can be complex and non-linear. To aid in the interpretation of the factors, factor rotation is often used. Factor rotation is a process that involves rotating the factors to simplify their interpretation.

The most common method for factor rotation is the Varimax rotation. The Varimax rotation produces factors that are uncorrelated with each other and have high factor loadings for a small number of variables. This method simplifies the interpretation of the factors by grouping variables that have high loadings on the same factor and reducing the complexity of the relationships between the factors and the observed variables.

For example, in a study of personality traits, the Varimax rotation might group variables related to extraversion, such as sociability, assertiveness, and talkativeness, into a single factor. Similarly, in a study of socio-economic status, the Varimax rotation might group variables related to education, such as years of schooling, academic achievement, and educational attainment, into a single factor.

The interpretation of the factors in factor analysis is complex and highly dependent on the specific dataset being analyzed and the application. Factor rotation, particularly the Varimax rotation, can aid in the interpretation of the factors by simplifying their interpretation and reducing the complexity of the relationships between the factors and the observed variables.

6.4 Applications of Factor Analysis

Factor analysis is a widely used statistical technique with many applications across various fields. In psychology, factor analysis is commonly used to identify underlying constructs that contribute to personality traits, intelligence, or cognitive abilities. For example, in a study of intelligence, factor analysis can be used to identify the cognitive abilities that contribute to overall intelligence, such as verbal comprehension, spatial reasoning, or working memory. Similarly, in a study of personality traits, factor analysis can be used to identify the underlying dimensions of personality, such as extraversion, agreeableness, or openness to experience.

In economics, factor analysis is used to identify the underlying factors that contribute to economic growth or financial performance. For example, in a study of economic growth, factor analysis can be used to identify the factors that contribute to growth, such as investment, exports, or technological innovation. Similarly, in a study of financial performance, factor analysis can be used to identify the factors that contribute to financial success, such as profitability, asset quality, or liquidity.

Factor analysis can also be used in market research to identify the underlying factors that contribute to consumer preferences or buying behavior. For example, in a study of consumer preferences for a particular product, factor analysis can be used to identify the factors that influence consumer decision-making, such as price, quality, convenience, or brand loyalty. Similarly, in a study of consumer buying behavior, factor analysis can be used to identify the underlying factors that influence purchasing decisions, such as social norms, personal values, or lifestyle factors.

In healthcare, factor analysis can be used to identify the underlying factors that contribute to patient outcomes or disease risk. For example, in a study of patient outcomes, factor analysis can be used to identify the factors that influence recovery or quality of life, such as treatment efficacy, patient satisfaction, or adherence to medication. Similarly, in a study of disease risk, factor analysis can be used to identify the factors that contribute to disease onset or progression, such as genetics, lifestyle factors, or environmental exposures.

Factor analysis is a versatile technique that can be used in many different applications across various fields. Its ability to identify underlying factors or constructs can help simplify complex datasets and provide insight into the key factors that contribute to observed patterns or outcomes.

6.5 Conclusion chapter 6

Factor analysis is a powerful technique in multivariate analysis that allows us to identify underlying factors that contribute to observed variables. By identifying the underlying factors, factor analysis can help simplify complex datasets and aid in interpretation and analysis. Factor analysis has many applications in various fields and can be used to analyze and interpret complex datasets. Understanding the principles and applications of factor analysis is essential for making informed decisions based on multivariate data.

Chapter 7: Cluster Analysis

Cluster analysis is a multivariate statistical technique used to group objects or individuals based on their similarity or dissimilarity in a set of variables. In this chapter, we will discuss the fundamentals of cluster analysis, including the different types of clustering methods, their applications, and their advantages and limitations.

7.1 Overview of Cluster Analysis

Cluster analysis is a powerful tool in data analysis that allows us to identify meaningful groups or clusters in a dataset. It is a useful technique in many fields, including marketing, biology, and social sciences. Cluster analysis can be used to identify groups of similar individuals, to identify patterns in customer preferences, or to identify different subtypes of a disease.

There are two main types of cluster analysis: hierarchical clustering and partitioning clustering. Hierarchical clustering involves creating a hierarchy of clusters, where smaller clusters are combined to form larger clusters. Partitioning clustering involves dividing the dataset into non-overlapping clusters based on a set of criteria.

7.2 Hierarchical Clustering Methods

Hierarchical clustering methods can be further divided into two types: agglomerative and divisive.

7.2.1 Agglomerative methods

Agglomerative clustering is a hierarchical clustering method that is widely used in various fields, including biology, social sciences, and computer science. This method starts with each observation as a separate cluster and then successively merges the most similar clusters until all observations are in a single cluster.

The merging process is guided by a distance metric, which measures the similarity or dissimilarity between clusters. There are various distance metrics that can be used, such as Euclidean distance, Manhattan distance, or cosine similarity, depending on the type of data being analyzed and the research question being addressed.

For example, in a study of gene expression data, agglomerative clustering can be used to identify clusters of genes with similar expression patterns. The distance metric in this case could be based on the correlation or cosine similarity between the expression profiles of different genes. The resulting clusters can then be analyzed to identify biological pathways or functions that are enriched in each cluster, providing insight into the underlying mechanisms of gene regulation and cellular processes.

In social sciences, agglomerative clustering can be used to identify groups of individuals with similar characteristics or behaviors. For example, in a study of consumer preferences, agglomerative clustering can be used to identify groups of consumers with similar preferences for certain products or features. The distance metric in this case could be based on the similarity of product ratings or purchase histories. The resulting clusters can then be used to develop targeted marketing strategies or personalized recommendations for different consumer groups.

In computer science, agglomerative clustering can be used for image segmentation, where pixels with similar colors or textures are grouped together to form objects or regions. The distance metric in this case could be based on the difference in color or texture features between pixels. The resulting clusters can then be used to identify objects or regions of interest in the image, providing a useful tool for image analysis and computer vision.

7.2.2 Divisive methods

Divisive clustering, also known as top-down clustering, is a clustering method that starts with all observations in a single cluster and then successively splits the clusters into smaller clusters until each observation is in a separate cluster. This method is the opposite of agglomerative clustering, which starts with each observation as a separate cluster and merges the most similar clusters until all observations are in a single cluster.

An example of the use of divisive clustering can be seen in the field of image segmentation. Divisive clustering can be used to segment an image into smaller regions that are visually distinct. For example, a grayscale image can be divided into several regions based on intensity levels, where each region contains pixels with similar intensity values.

The algorithm starts by treating the entire image as a single cluster and then splits it into smaller clusters based on the intensity values of the pixels. The process is repeated until each pixel is in a separate cluster. This results in a segmentation of the image into regions that are visually distinct based on their intensity values.

Another example of the use of divisive clustering can be seen in the field of market research. Divisive clustering can be used to segment customers into smaller groups based on their behavior or preferences. For example, a company can use divisive clustering to identify groups of customers who have similar purchasing habits or preferences. The algorithm starts with all customers in a single cluster and then splits the cluster into smaller clusters based on their purchasing habits or preferences. The process is repeated until each customer is in a separate cluster. This results in a segmentation of customers into groups that are similar in terms of their purchasing habits or preferences. This information can be used by the company to develop targeted marketing strategies for each group of customers.

Hierarchical clustering methods are useful for exploring the structure of the data and visualizing the clustering results. However, they can be computationally intensive, especially for large datasets, and can produce dendrograms that are difficult to interpret.

7.3 Partitioning Methods

Partitioning methods involve dividing the dataset into a specified number of clusters based on a set of criteria. The most common partitioning method is K-means clustering, which involves iteratively assigning observations to the nearest cluster centroid and then recalculating the centroids.

7.3.1 K-means Clustering

K-means clustering is a widely used partitioning method in unsupervised machine learning that involves iteratively assigning observations to the nearest cluster centroid and then recalculating the centroids. The goal of K-means clustering is to partition the observations into K clusters, where each observation belongs to the cluster whose centroid is closest to it.

An example of the use of K-means clustering can be seen in the field of customer segmentation. K-means clustering can be used to segment customers into different groups based on their purchasing behavior, demographics, or other relevant variables. For example, a retail store might use K-means clustering to segment their customers based on their past purchasing behavior. The algorithm would begin by randomly assigning each customer to one of K clusters and then iteratively updating the cluster centroids based on the mean value of the observations in each cluster. The algorithm would continue until the clusters no longer changed or until a maximum number of iterations was reached.

Another example of the use of K-means clustering can be seen in the field of image compression. K-means clustering can be used to reduce the number of colors in an image by grouping similar colors together. For example, an image with 256 different colors could be compressed to an image with 16 colors by using K-means clustering to group similar colors into clusters. The algorithm would begin by randomly selecting K colors as the initial centroids and then iteratively updating the centroids and assigning each pixel to the nearest centroid. The final image would be reconstructed by replacing each pixel with the centroid color to which it was assigned.

K-means clustering is a simple and efficient method that can be used for a wide range of applications, including data clustering, image processing, and customer segmentation. However, it is important to note that the results of K-means clustering can be sensitive to the initial choice of centroids and the number of clusters chosen, and may not always lead to the optimal clustering solution.

7.4 Applications of Cluster Analysis

Cluster analysis has many applications in various fields. In marketing, cluster analysis can be used to identify different groups of customers based on their preferences and behavior. In biology, cluster analysis can be used to identify different subtypes of a disease based on genetic markers. In social sciences, cluster analysis can be used to identify different groups of individuals based on their attitudes and beliefs.

Cluster analysis can also be used in image analysis to identify different regions of an image based on their texture or color. In transportation, cluster analysis can be used to identify different patterns of travel behavior based on the characteristics of the trips.

7.5 Conclusion chapter 7

Cluster analysis is a powerful technique in multivariate analysis that allows us to identify meaningful groups or clusters in a dataset. By identifying the groups, cluster analysis can help simplify complex datasets and aid in interpretation and analysis. Cluster analysis has many applications in various fields and can be used to analyze and interpret complex datasets. Understanding the principles and applications of cluster analysis is essential for making informed decisions based on multivariate data.

Chapter 8: Structural Equation Modeling

Structural Equation Modeling (SEM) is a powerful statistical technique that is widely used to analyze complex relationships between variables. SEM is a combination of factor analysis and regression analysis, where it allows researchers to test theoretical models and understand the causal relationships between variables. In this chapter, we will provide a comprehensive overview of SEM, its uses, and how it is applied in different fields.

8.1 Overview of Structural Equation Modeling

Structural Equation Modeling is a statistical method that allows the estimation of complex relationships among multiple variables. It is a versatile technique that enables researchers to test a wide range of models, including path analysis, confirmatory factor analysis, and latent growth models. SEM can be used to test and estimate both linear and nonlinear relationships among variables.

In structural equation modeling, the relationships between variables are represented as a set of equations. The equations specify how each variable is related to other variables in the model, either directly or indirectly. The model is then tested to determine whether the relationships among variables are significant and whether the model fits the data well.

8.2 Model Specification and Estimation

Model specification is a critical component of SEM. Researchers must have a clear idea of the variables they wish to include in their model and the relationships among them. SEM models can be specified using either a structural or a measurement model.

Structural models specify the relationships among variables in a system. These models can be either causal or non-causal, depending on the researcher's hypotheses. In causal models, researchers specify the direction of causality, while in non-causal models, they do not.

Measurement models specify the relationships between variables and their associated measurement instruments. They are used to estimate the reliability and validity of a set of measures used to assess a particular construct.

Once the model has been specified, it is estimated using maximum likelihood estimation or another suitable method. Maximum likelihood estimation is a common method used to estimate model parameters in SEM. The goal of maximum likelihood estimation is to find the values of the model parameters that maximize the likelihood of obtaining the observed data.

8.3 Goodness-of-Fit Measures

Goodness-of-fit measures are used to assess the degree to which the estimated model fits the observed data. The most common goodness-of-fit measures used in SEM are the chi-square statistic, the Comparative

Fit Index (CFI), and the Root Mean Square Error of Approximation (RMSEA).

The chi-square statistic is a measure of the difference between the observed data and the model predictions. It is used to test the null hypothesis that the model fits the data well. A significant chi-square value indicates a poor fit between the model and the data.

The Comparative Fit Index (CFI) is a measure of how well the model fits the data, relative to a null model. The CFI ranges from 0 to 1, with values closer to 1 indicating a better fit.

The Root Mean Square Error of Approximation (RMSEA) is a measure of the discrepancy between the model and the data. It estimates the average difference between the observed and predicted data, with lower values indicating a better fit.

8.4 Applications of Structural Equation Modeling

Structural Equation Modeling is a widely used statistical technique that has a wide range of applications in various fields. SEM has been used in social sciences, marketing, finance, education, and healthcare, among other fields.

In social sciences, SEM has been used to study the relationships between various psychological constructs, such as self-esteem, depression, and anxiety. It has also been used to examine the effects of social support on health outcomes, such as heart disease.

In marketing, SEM has been used to study consumer behavior, such as the effects of advertising on brand loyalty and customer satisfaction.

In finance, SEM has been used to study the relationships between various financial indicators, such as stock prices, interest rates, and inflation rates.

Structural equation modeling (SEM) is a statistical tool that is widely used in finance to analyze the complex relationships between various financial indicators such as stock prices, interest rates, and inflation rates. SEM enables researchers to build and test models that specify the relationships between multiple variables, which is particularly useful in finance where there are often multiple variables that can impact each other.

One of the key benefits of SEM is that it allows researchers to test causal relationships between variables. For example, researchers can use SEM to investigate whether changes in interest rates cause changes in stock prices or whether changes in inflation rates cause changes in interest rates. By examining the causal relationships between variables, researchers can gain a deeper understanding of the factors that drive financial markets and make more informed investment decisions.

Another advantage of SEM is that it can be used to model complex relationships between variables. For example, researchers can use SEM to investigate whether stock prices are influenced by a combination of macroeconomic factors such as interest rates, inflation rates, and economic growth, as well as company-specific factors such as earnings and dividends. By modeling these complex relationships, researchers can gain a more nuanced understanding of the factors that drive stock prices and develop more accurate forecasting models.

In addition to these benefits, SEM also allows researchers to test the validity of their models using goodness-of-fit measures. These measures enable researchers to assess how well their models fit the data and identify areas where the model may need to be revised or refined.

Overall, SEM is a powerful tool that has been widely used in finance to analyze the complex relationships between various financial indicators. By enabling researchers to model causal relationships and complex interactions between variables, SEM has helped to advance our understanding of financial markets and inform investment decisions.

Chapter 9: Time Series Analysis

Time series analysis is a statistical method used to analyze data that is collected over time. This method is commonly used in fields such as economics, finance, and engineering to identify trends, patterns, and relationships in time-dependent data. Time series analysis is useful in forecasting future values, understanding the underlying factors that influence a time series, and identifying patterns or anomalies in the data.

Autoregressive models, moving average models, and ARIMA models are some of the most common time series models used in statistical analysis. Autoregressive models are based on the idea that a value in a time series can be predicted based on the previous values in the same series. Moving average models use the past errors in the time series to predict future values. ARIMA models combine the autoregressive and moving average models, along with differencing to remove any non-stationary characteristics of the time series.

One of the main applications of time series analysis is in forecasting. Time series models can be used to predict future values based on historical data. This is particularly useful in fields such as finance and economics, where accurate forecasts of economic indicators such as stock prices, interest rates, and GDP can inform investment decisions and economic policy.

Another application of time series analysis is in identifying trends and patterns in the data. For example, a time series analysis of stock prices may reveal seasonal patterns or cyclical trends that can inform investment strategies. Similarly, a time series analysis of weather data can help to identify long-term climate trends or predict weather patterns.

Time series analysis is also useful in identifying anomalies or outliers in the data. These anomalies may represent significant events or changes in the underlying factors that influence the time series. By identifying and understanding these anomalies, researchers can gain insight into the factors that drive the time series and develop more accurate forecasting models.

Overall, time series analysis is a powerful statistical method that is widely used in fields such as finance, economics, and engineering. By identifying trends, patterns, and anomalies in time-dependent data, time series analysis enables researchers to make more informed decisions and develop more accurate forecasting models.

Chapter 10: Multivariate Methods in Big Data Analytics

In recent years, the explosive growth of big data has led to an increased need for multivariate methods that can effectively analyze large, complex datasets. Big data sets are often characterized by their large number of variables and observations, making traditional statistical methods impractical or inefficient. Multivariate methods, however, offer a powerful solution to these challenges, allowing analysts to identify patterns and relationships within large datasets, and extract insights that can inform decision-making and drive innovation.

One of the main challenges of analyzing large multivariate datasets is that traditional statistical methods may not be scalable to handle the size and complexity of the data. For example, traditional regression analysis may be computationally expensive or impractical when dealing with thousands or even millions of variables. Additionally, big data sets often contain missing or incomplete data, which can further complicate analysis.

To overcome these challenges, many researchers have turned to machine learning methods, which are specifically designed to analyze large and complex datasets. Machine learning methods involve training algorithms on large datasets to learn patterns and relationships, and then using these algorithms to make predictions or classify new data. Some common machine learning methods used in multivariate analysis include decision trees, neural networks, and support vector machines.

Another important application of machine learning in multivariate analysis is in data clustering and classification. Clustering algorithms can group similar observations together based on their similarities, while classification algorithms can predict the class or category of new observations based on their features. These methods can be used in a variety of fields, from marketing to healthcare, to identify patterns and relationships within complex datasets and make informed decisions based on those insights.

In conclusion, multivariate methods are critical to analyzing big data sets, enabling researchers and analysts to identify patterns and relationships within complex datasets, and extract insights that can drive innovation and inform decision-making. Machine learning methods in particular have emerged as a powerful tool in multivariate analysis, offering scalable solutions to the challenges posed by large and complex datasets. By leveraging these methods, researchers and analysts can unlock new insights and opportunities within the ever-growing field of big data analytics.

Chapter 11: Multidimensional scaling

Multidimensional scaling (MDS) is a statistical technique used to analyze the similarities and differences between objects or individuals based on multiple variables. It is used to visualize complex relationships and patterns in multivariate data and is widely used in fields such as psychology, marketing, and geography.

MDS involves taking a set of measurements or variables for a set of objects or individuals and creating a low-dimensional representation of those objects that preserves the similarity relationships between them. The goal is to create a visualization of the data that is easy to interpret and provides insight into the underlying relationships between the variables and objects.

There are two main types of MDS: metric and non-metric. Metric MDS assumes that the distance between objects in the low-dimensional space is a metric, meaning that it satisfies the triangle inequality. Non-metric MDS does not make this assumption, but still attempts to preserve the similarity relationships between objects.

To perform MDS, the first step is to calculate the distance matrix between objects based on the variables of interest. This distance matrix can be calculated using a variety of distance measures, such as Euclidean distance, Manhattan distance, or correlation distance. Once the distance matrix is calculated, the MDS algorithm works to place the objects in a low-dimensional space while preserving the distances between them.

The resulting visualization is often in the form of a scatterplot, where each point represents an object and the distance between points represents the similarity or dissimilarity between the objects. MDS can also be used to identify clusters of similar objects, as well as to identify variables that are most important in distinguishing between different groups of objects.

MDS has numerous applications in various fields. In psychology, it is often used to study the perceived similarity between stimuli such as faces, colors, or words. In marketing, it can be used to analyze the similarities and differences between different brands or products. In geography, it can be used to study the spatial relationships between different locations or regions.

In conclusion, multidimensional scaling is a powerful statistical technique used to visualize complex relationships and patterns in multivariate data. It allows for the creation of a low-dimensional representation of the data that preserves the similarity relationships between objects. MDS has numerous applications in various fields and is a valuable tool for understanding the relationships between variables and objects in multivariate data.

Chapter 12: Bayesian Methods for Multivariate Analysis

In recent years, Bayesian methods have gained immense popularity in various fields of research, including multivariate analysis. Bayesian methods offer a flexible and powerful framework for modeling complex multivariate data and allow for the incorporation of prior knowledge and uncertainty in the analysis. In this chapter, we will provide an overview of Bayesian methods for multivariate analysis, including Bayesian inference, Bayesian regression, Bayesian factor analysis, and Bayesian network analysis.

12.1.1 Bayesian Inference

Bayesian inference is a powerful approach to statistical inference that allows us to quantify uncertainty in our estimates. Bayesian inference involves updating our prior knowledge of a parameter or a set of parameters based on the observed data, resulting in a posterior distribution. The posterior distribution represents the updated knowledge about the parameter(s) of interest, taking into account the prior knowledge and the observed data. The posterior distribution can be used to estimate point estimates (e.g., mean, median) and credible intervals for the parameter(s) of interest.

12.1.2 Bayesian Regression

Bayesian regression is a flexible and powerful approach to regression analysis that allows us to incorporate prior knowledge and uncertainty in the analysis. Bayesian regression models allow us to estimate the parameters of a regression model using a posterior distribution, which incorporates the prior distribution and the likelihood function of the data. Bayesian regression models can be used for both linear and non-linear regression problems and can handle both continuous and categorical predictors.

12.1.3 Bayesian Factor Analysis

Bayesian factor analysis is a popular method for dimensionality reduction and variable selection in multivariate data. Bayesian factor analysis allows us to identify a small number of underlying factors that explain the observed correlations between variables. Bayesian factor analysis models can be used to estimate the factor loadings, factor scores, and residual variances. Bayesian factor analysis models can also be extended to include prior distributions on the factor loadings and to allow for missing data.

12.1.4 Bayesian Network Analysis

Bayesian network analysis is a powerful method for modeling complex multivariate relationships. Bayesian networks are graphical models that represent the conditional dependencies between variables in a probabilistic manner. Bayesian network analysis can be used to identify the most influential variables, estimate the strength of the relationships between variables, and perform prediction and classification tasks.

12.2 Conclusion chapter 12

Bayesian methods provide a flexible and powerful framework for modeling complex multivariate data. Bayesian methods allow us to incorporate prior knowledge and uncertainty in the analysis and provide a natural way to deal with missing data and variable selection problems. Bayesian methods have found applications in various fields of research, including finance, marketing, biology, and engineering. The use of Bayesian methods in multivariate analysis is expected to increase in the coming years as the size and complexity of datasets continue to grow.

Chapter 13: Multivariate outlier detection and robust methods

13.1 Introduction

Multivariate analysis deals with data sets that have more than one variable. The methods used in multivariate analysis are designed to examine the relationships between multiple variables, which can be difficult to uncover using univariate or bivariate techniques. One common problem in multivariate analysis is the presence of outliers, which can distort the results of statistical analysis. Outliers can be caused by measurement errors, data entry errors, or the presence of unusual cases in the data set. Robust methods are designed to detect and handle outliers in multivariate analysis.

13.2 Multivariate Outlier Detection

Multivariate outlier detection involves identifying observations that are unusual in the context of the entire data set, rather than just in one variable. There are several methods that can be used to detect multivariate outliers, including:

1. Mahalanobis distance: This method calculates the distance between each observation and the center of the data set, taking into account the correlations between variables. Observations that have a large Mahalanobis distance are considered outliers.

2. Robust Mahalanobis distance: This method is similar to the Mahalanobis distance, but uses robust estimators of the mean and covariance matrix to handle outliers.

3. Cook's distance: This method measures the influence of each observation on the regression coefficients. Observations that have a large Cook's distance are considered influential and may be outliers.

4. Distance-based methods: These methods use a distance metric to measure the similarity between observations. Observations that are far away from the rest of the data set may be outliers.

13.3 Robust Methods

Robust methods are designed to handle outliers in multivariate analysis. These methods are less sensitive to the presence of outliers than traditional methods, which means that they produce more reliable results in the presence of outliers. Some commonly used robust methods include:

1. Robust regression: This method uses robust estimators of the mean and covariance matrix to handle outliers in the data set. It is similar to traditional regression, but is less sensitive to outliers.

2. Robust principal component analysis: This method is used to identify the principal components of the data set, while handling

outliers. It is similar to traditional principal component analysis, but is less sensitive to outliers.

3. Robust covariance matrix estimation: This method is used to estimate the covariance matrix of the data set, while handling outliers. It is similar to traditional covariance matrix estimation, but is less sensitive to outliers.

13.4 Conclusion chapter 13

Multivariate outlier detection and robust methods are important tools for analyzing multivariate data sets. They can help to identify outliers and handle them in a way that produces more reliable results. Outliers can be caused by a variety of factors, including measurement errors, data entry errors, and the presence of unusual cases in the data set. Robust methods are designed to handle these outliers, making them an important tool for multivariate analysis.

Chapter 14: Nonlinear multivariate analysis techniques

14.1 Introduction

Multivariate analysis is a powerful tool for exploring relationships between multiple variables. In many cases, linear methods are sufficient for analyzing multivariate data. However, there are situations in which the relationships between variables are nonlinear, and linear methods may not capture the full complexity of the data. In such cases, nonlinear multivariate analysis techniques may be used.

14.2 Nonlinear Multivariate Analysis Techniques

A. Kernel Methods

Kernel methods are a family of nonlinear techniques that use a kernel function to transform data into a higher-dimensional space where linear techniques can be applied. The kernel function maps the original data into a new space, in which a linear boundary can be found to separate the data. Kernel methods are commonly used for classification and regression problems.

B. Neural Networks

Neural networks are a class of nonlinear models that consist of interconnected nodes, or neurons. The neurons receive inputs and produce outputs that are passed to other neurons in the network. The connections between neurons have weights that are adjusted during training to optimize the network's performance on a given task. Neural networks are used for a wide range of applications, including pattern recognition, classification, and prediction.

C. Decision Trees

Decision trees are a type of nonlinear model that uses a tree-like structure to represent a sequence of decisions that lead to a particular outcome. Each node in the tree represents a decision based on the value of a particular variable, and each branch represents one of the possible outcomes of that decision. Decision trees can be used for classification, regression, and other types of data analysis.

D. Support Vector Machines

Support vector machines (SVMs) are a type of nonlinear model that use a kernel function to map data into a higher-dimensional space, in which a linear boundary can be found to separate the data. SVMs are particularly useful for classification problems, where they are often used to find a decision boundary that maximizes the margin between different classes of data.

E. Nonlinear Dimensionality Reduction

Nonlinear dimensionality reduction techniques are used to find a low-dimensional representation of high-dimensional data, while preserving the essential characteristics of the data. Nonlinear dimensionality reduction techniques include manifold learning, kernel PCA, and locally linear embedding.

14.3 Advantages and Limitations

Nonlinear multivariate analysis techniques offer several advantages over linear methods. They are capable of capturing complex relationships between variables, and can provide more accurate predictions and classifications. However, they can also be computationally expensive and may require large amounts of data for training. In addition, they can be more difficult to interpret than linear methods, and may require specialized knowledge and expertise to use effectively.

14.4 Applications

Nonlinear multivariate analysis techniques have a wide range of applications in various fields, including:

- Image and Signal Processing: Nonlinear multivariate techniques are widely used in image and signal processing applications, such as image recognition, pattern recognition, and speech recognition. These techniques help to extract relevant features from images and signals, which can be used for classification, clustering, and other tasks.

- Natural Language Processing: Nonlinear multivariate analysis techniques are also useful in natural language processing tasks, such as sentiment analysis, text classification, and topic modeling. These techniques help to extract relevant features from text data, which can be used to build predictive models.

- Bioinformatics: Nonlinear multivariate analysis techniques are used in bioinformatics to analyze complex biological data, such as gene expression data, protein-protein interaction networks, and metabolic pathways. These techniques help to identify relationships between different variables and to build predictive models for various biological processes.

- Finance: Nonlinear multivariate analysis techniques are used in finance to analyze complex financial data, such as stock prices, interest rates, and currency exchange rates. These techniques help to identify relationships between different variables and to build predictive models for various financial processes.

- Marketing: Nonlinear multivariate analysis techniques are used in marketing to analyze customer data and to build predictive models for various marketing processes, such as customer segmentation, product recommendation, and market forecasting.

- Neuroscience: Nonlinear multivariate analysis techniques are used in neuroscience to analyze brain activity data, such as electroencephalogram (EEG) data, functional magnetic resonance imaging (fMRI) data, and single-cell recordings. These techniques help to identify relationships between different brain regions and to build predictive models for various cognitive processes.

- Environmental Science: Nonlinear multivariate analysis techniques are used in environmental science to analyze complex environmental data, such as climate data, water quality data, and air pollution data. These techniques help to identify relationships between different environmental variables and to build predictive models for various environmental processes.

- Robotics: Nonlinear multivariate analysis techniques are used in robotics to analyze sensor data, such as lidar and camera data, and to build predictive models for various robotic processes, such as navigation, object detection, and grasping. These techniques help to identify relationships between different sensor measurements and to build more accurate and robust robotic systems.

14.5 Conclusion chapter 14

Nonlinear multivariate analysis techniques are powerful tools for exploring complex relationships between multiple variables. They offer several advantages over linear methods, but also have some limitations. The choice of technique depends on the specific problem being addressed, as well as the available data and computational resources. As the field of multivariate analysis continues to evolve, nonlinear techniques are likely to play an increasingly important role in data analysis and interpretation.

Chapter 15: Multivariate Analysis of Repeated Measures Data

In many research studies, the same variable is measured multiple times over a period of time, or in different experimental conditions. Such data is called repeated measures data. Analyzing such data requires a multivariate approach since the observations are no longer independent. This chapter will provide an overview of the multivariate techniques used in the analysis of repeated measures data.

15.1 Overview of Repeated Measures Data

Repeated measures data refer to data where each subject is measured multiple times. The observations are therefore not independent, and traditional univariate methods may not be appropriate. This type of data is common in longitudinal studies, clinical trials, and experimental designs with multiple treatments or conditions.

15.2 Multivariate Techniques for Repeated Measures Data

There are several multivariate techniques for analyzing repeated measures data, including:

1. Multivariate Analysis of Variance (MANOVA): MANOVA is used to test whether there is a significant difference in means across multiple dependent variables. MANOVA can be extended to

repeated measures data by adding the within-subjects factors to the model.

2. Multivariate Analysis of Covariance (MANCOVA): MANCOVA is similar to MANOVA, but it includes one or more continuous covariates in the model. MANCOVA can be extended to repeated measures data by including the within-subjects factors and covariates in the model.

3. Linear Mixed Models (LMM): LMM is a flexible method for analyzing repeated measures data that accounts for both within-subjects and between-subjects variability. LMM can handle unbalanced data, missing values, and different covariance structures.

4. Generalized Estimating Equations (GEE): GEE is a robust and flexible method for analyzing repeated measures data. GEE does not require any assumptions about the distribution of the response variable or the covariance structure, making it a good choice for non-normal and non-linear data.

5. Structural Equation Modeling (SEM): SEM is a powerful method for analyzing complex relationships among multiple variables. SEM can be used to model the within-subjects and between-subjects factors and their interactions.

15.3 Applications of Multivariate Techniques for Repeated Measures Data

Multivariate techniques for repeated measures data have been applied in many fields, including:

1. Longitudinal studies: Multivariate techniques are commonly used in longitudinal studies to investigate changes over time in multiple variables, such as cognitive functioning, physical health, and quality of life.

2. Clinical trials: Multivariate techniques can be used to analyze clinical trial data with repeated measures of multiple outcomes, such as pain, function, and quality of life.

3. Psychology: Multivariate techniques are used in psychology research to investigate the effects of different treatments or interventions on multiple psychological measures, such as depression, anxiety, and stress.

4. Education: Multivariate techniques are used in education research to investigate the effects of educational interventions on multiple outcomes, such as academic achievement, motivation, and attitudes towards learning.

15.4 Conclusion chapter 15

Repeated measures data require a multivariate approach to account for the dependence among observations. Several multivariate techniques are available for analyzing repeated measures data, including MANOVA, MANCOVA, LMM, GEE, and SEM. These methods can be applied in various fields, including longitudinal studies, clinical trials, psychology, and education research. Choosing the appropriate method depends on the research question, the nature of the data, and the assumptions underlying the method.

Chapter 16: Network Analysis in Multivariate Data

16.1 Introduction

Multivariate data analysis deals with multiple variables that are interrelated and interact with each other. Network analysis is a tool used to study the relationships between variables in a multivariate dataset. Network analysis helps to visualize and understand the complex relationships between variables in a dataset. In this chapter, we will discuss the use of network analysis in multivariate data analysis.

16.2 Overview of Network

Analysis Network analysis is a method of analyzing complex systems using a network of interconnected elements. In a network, each element is represented as a node, and the relationships between the nodes are represented as edges. In multivariate data analysis, network analysis is used to study the relationships between variables. The variables are represented as nodes, and the relationships between the variables are represented as edges.

1. Types of Networks

   There are two types of networks used in multivariate data analysis: correlation networks and partial correlation networks. Correlation networks are constructed using the correlation matrix of the variables in the dataset. Partial correlation networks, on the other hand, are constructed using the partial correlation matrix of the variables in the dataset. Partial correlation networks are used to study the relationships between variables after removing the effects of other variables.

2. Network Visualization

Network visualization is an important tool in network analysis. It helps to visualize the relationships between variables in a network. There are several methods used for network visualization, including spring-embedded layouts, circular layouts, and force-directed layouts.

3. Network Metrics

Network metrics are used to quantify the properties of a network. Some of the commonly used network metrics include degree centrality, betweenness centrality, and clustering coefficient. Degree centrality measures the number of edges connected to a node. Betweenness centrality measures the importance of a node in connecting different parts of the network. Clustering coefficient measures the degree to which nodes tend to cluster together.

4. Applications of Network Analysis in Multivariate Data Analysis
Network analysis has several applications in multivariate data analysis. It is used in gene expression analysis to study the relationships between genes. It is also used in social network analysis to study the relationships between individuals in a social network. In finance, network analysis is used to study the

relationships between financial assets. In epidemiology, network analysis is used to study the spread of infectious diseases.

16.3 Conclusion chapter 16

Network analysis is a powerful tool for studying the relationships between variables in a multivariate dataset. It helps to visualize and understand the complex relationships between variables. Network analysis has several applications in different fields, including gene expression analysis, social network analysis, finance, and epidemiology.

Chapter 17 Time-Frequency Analysis of Multivariate Time Series Data

17. 1 Introduction

Time series data is a sequence of observations taken at equally spaced time intervals, and multivariate time series data refers to a collection of such sequences, each representing a different variable. These data sets are commonly used in various fields, such as finance, economics, meteorology, neuroscience, and signal processing. One of the challenges of analyzing multivariate time series data is that the relationship between variables can change over time, which makes traditional time series methods less effective. Time-frequency analysis is a powerful technique that can reveal how the power of a signal is distributed across both time and frequency domains. In this chapter, we will explore various methods of time-frequency analysis of multivariate time series data.

17.2 Overview of Time-Frequency Analysis

Time-frequency analysis is a technique that analyzes the time-varying frequency content of a signal. The goal is to identify how the power of a signal is distributed over time and frequency domains. Time-frequency analysis is particularly useful for analyzing non-stationary signals, where the relationship between the variables changes over time. There are several methods of time-frequency analysis, including wavelet transforms, spectrograms, and scalograms.

1. Wavelet Transforms

   Wavelet transforms are a popular method for time-frequency analysis of multivariate time series data. The wavelet transform decomposes the signal into a set of time-frequency components using a set of basis functions known as wavelets. The resulting time-frequency representation can reveal changes in the frequency content of the signal over time. Wavelet transforms have been used in various fields, such as finance, meteorology, and neuroscience, to analyze multivariate time series data.

2. Spectrograms

   Spectrograms are another method of time-frequency analysis that provides a time-varying spectral representation of a signal. Spectrograms are obtained by taking the Fourier transform of short segments of the signal and plotting the resulting power spectrum as a function of time. The resulting spectrogram can reveal how the power of the signal is distributed across different frequencies over time. Spectrograms have been used in fields such as speech recognition and music analysis.

3. Scalograms

   Scalograms are a type of time-frequency analysis that uses a two-dimensional representation of the wavelet transform coefficients. The scalogram is a color-coded representation of the wavelet coefficients, where the color represents the magnitude of the coefficient. The resulting scalogram can reveal how the power of the signal is distributed across both time and frequency domains.

4. Applications of Time-Frequency Analysis

    Time-frequency analysis has a wide range of applications in various fields, including finance, economics, meteorology, neuroscience, and signal processing. For example, in finance, time-frequency analysis can be used to study the relationship between various financial variables, such as stock prices, interest rates, and inflation rates, and how this relationship changes over time. In neuroscience, time-frequency analysis can be used to analyze the relationship between brain waves and behavior, and how this relationship changes over time. In meteorology, time-frequency analysis can be used to study the relationship between weather variables, such as temperature and humidity, and how this relationship changes over time.

17.3 Conclusion

Time-frequency analysis is a powerful technique for analyzing multivariate time series data. It can reveal how the power of a signal is distributed across both time and frequency domains and is particularly useful for analyzing non-stationary signals where the relationship between variables changes over time. Wavelet transforms, spectrograms, and scalograms are all popular methods of time-frequency analysis. Time-frequency analysis has a wide range of applications in various fields, including finance, economics, meteorology, neuroscience, and signal processing.

Chapter 18: Multivariate Meta-Analysis

18.1 Introduction

Meta-analysis is a statistical technique that combines results from multiple studies to increase the precision and validity of estimates. It has become a popular tool in evidence-based research, particularly in the medical and social sciences. In traditional meta-analysis, studies are typically analyzed one outcome at a time, assuming that each outcome is independent. However, in many cases, studies report multiple outcomes that are related to each other. In such cases, multivariate meta-analysis can be used to simultaneously analyze multiple outcomes.

18.2 Definition of Multivariate Meta-Analysis

Multivariate meta-analysis is a statistical technique that allows for the simultaneous analysis of multiple outcomes in meta-analysis. It is used when studies report multiple outcomes that are correlated with each other. The technique takes into account the correlations between outcomes and provides a more accurate estimate of the overall effect size than traditional meta-analysis.

18.3 Multivariate Meta-Analysis Methods

There are several methods available for conducting multivariate meta-analysis, including the maximum likelihood estimation (MLE) and restricted maximum likelihood estimation (REML) methods. The MLE method assumes that the effect sizes for each outcome are normally distributed, whereas the REML method assumes that the effect sizes are a random sample from a larger population of effect sizes.

In addition to the MLE and REML methods, there are several other methods for conducting multivariate meta-analysis, such as the Bayesian hierarchical model, which allows for the incorporation of prior knowledge into the analysis.

18.4 Importance of Multivariate Meta-Analysis

Multivariate meta-analysis has several advantages over traditional meta-analysis. It allows for the simultaneous analysis of multiple outcomes, which can lead to more precise estimates of the overall effect size. In addition, it can provide a better understanding of the relationships between different outcomes, and can identify potential moderators of the overall effect.

18.5 Applications of Multivariate Meta-Analysis

Multivariate meta-analysis has been used in a variety of fields, including medicine, psychology, and education. It has been used to analyze multiple outcomes in studies of treatment effectiveness, to examine the relationships between different outcomes, and to identify potential moderators of treatment effects.

For example, a multivariate meta-analysis was conducted to investigate the effectiveness of cognitive-behavioral therapy (CBT) for anxiety disorders. The analysis included multiple outcomes, such as anxiety symptoms, depression symptoms, and quality of life. The results showed that CBT was effective for all outcomes, but the effect sizes were larger for anxiety symptoms than for the other outcomes.

Conclusion chapter 18

Multivariate meta-analysis is a powerful tool for analyzing multiple outcomes in meta-analysis. It allows for the simultaneous analysis of multiple outcomes, which can lead to more precise estimates of the overall effect size. It can also provide a better understanding of the relationships between different outcomes and identify potential moderators of the overall effect. Multivariate meta-analysis has important applications in medicine, psychology, and education, and its use is likely to increase as researchers continue to recognize its benefits.

Chapter 19: Multivariate Visualization Techniques

19.1 Introduction

Multivariate analysis involves the study of multiple variables simultaneously, making it challenging to visualize the data in a way that is both informative and aesthetically pleasing. Multivariate visualization techniques allow analysts to effectively represent and communicate the complex relationships between multiple variables. This chapter provides an overview of multivariate visualization techniques, including scatterplot matrices, parallel coordinates, and heatmaps.

19.2 Scatterplot Matrices

Scatterplot matrices are a type of multivariate visualization technique that allows for the simultaneous display of pairwise relationships between multiple variables. In a scatterplot matrix, each variable is plotted against every other variable, resulting in a grid of scatterplots. Scatterplot matrices are useful for identifying patterns and relationships between variables. They can also be used to detect outliers and clusters within the data.

19.3 Parallel Coordinates

Parallel coordinates are a multivariate visualization technique that can be used to visualize high-dimensional data. In parallel coordinates, each variable is represented by a vertical axis, and the data points are connected by a line that passes through each axis. Parallel coordinates are useful for identifying patterns and relationships between variables. They can also be used to identify outliers and clusters within the data.

19.4 Heatmaps

Heatmaps are a type of multivariate visualization technique that can be used to represent the relationships between multiple variables in a matrix format. Heatmaps use color-coding to represent the magnitude of the relationship between each variable pair. The color-coding can be used to identify patterns and relationships within the data, and can also be used to identify outliers and clusters.

19.5 Other Techniques

There are many other multivariate visualization techniques that can be used to effectively represent complex data relationships, including:

- Trellis plots: a series of small plots that show the relationship between variables in different subsets of the data

- 3D scatterplots: a scatterplot with a third dimension added, allowing for the visualization of three variables simultaneously

- Glyph plots: a multivariate visualization technique that uses symbols to represent the values of multiple variables

19.6 Applications of Multivariate Visualization Techniques

Multivariate visualization techniques are useful in a variety of fields, including finance, biology, and social sciences. For example, in finance, multivariate visualization techniques can be used to visualize the relationships between multiple financial indicators, such as stock prices, interest rates, and inflation rates. In biology, multivariate visualization techniques can be used to visualize the relationships between multiple gene expression levels, and in the social sciences, multivariate visualization techniques can be used to visualize the relationships between multiple demographic variables.

Conclusion chapter 19

Multivariate visualization techniques play a crucial role in the analysis and interpretation of complex data relationships. Scatterplot matrices, parallel coordinates, and heatmaps are just a few examples of the many multivariate visualization techniques that can be used to represent complex data relationships. By effectively representing and communicating the relationships between multiple variables, multivariate visualization techniques can help analysts gain insights and make informed decisions based on the data.

Chapter 20: Conclusion

Multivariate analysis is a powerful and versatile tool for exploring and understanding complex relationships among multiple variables. This comprehensive guide has covered a wide range of topics in multivariate analysis, from the basic principles of multivariate distributions and descriptive statistics to advanced techniques like Bayesian methods and time-frequency analysis of multivariate time series data.

One of the most important takeaways from this guide is the importance of understanding the context and purpose of any multivariate analysis. Different types of data and research questions may require different techniques and methods, and it is essential to choose the most appropriate approach for the specific task at hand.

Another key message is that multivariate analysis is not a static field, but rather a rapidly evolving area of research. New methods and techniques are constantly being developed and refined, and there are many exciting opportunities for future research in this area.

Despite the challenges and complexities of multivariate analysis, the potential benefits are vast. By using these techniques to uncover underlying patterns and relationships, we can gain new insights and make better decisions in a wide range of fields, from finance and economics to healthcare and social sciences.

In conclusion, multivariate analysis is an essential tool for modern data analysis, offering unparalleled insights into complex relationships among multiple variables. With a solid understanding of the principles and techniques covered in this guide, researchers can leverage the power of multivariate analysis to unlock new knowledge and drive innovation in their respective fields.

Chapter 21: references and further reading

Books

1. "Multivariate Data Analysis" by Hair, Anderson, Tatham, and Black

2. "Applied Multivariate Statistical Analysis" by Richard A. Johnson and Dean W. Wichern

3. "The Elements of Statistical Learning: Data Mining, Inference, and Prediction" by Trevor Hastie, Robert Tibshirani, and Jerome Friedman

4. "Multivariate Analysis Techniques in Social Science Research: From Problem to Analysis" by Jacques A. Hagenaars and Allan L. McCutcheon

5. "Principal Component Analysis" by I.T. Jolliffe

Articles

1. "Multivariate Analysis" by William Revelle

2. "A Tutorial on Principal Component Analysis" by Jonathon Shlens

3. "A Gentle Introduction to Multivariate Analysis" by David Hood